geog.2

5th edition

workbook

<justin woolliscroft>
<katy patchwood>

Name:

Class:

OXFORD

Great Clarendon Street, Oxford, OX2 6DP, United Kingdom

Oxford University Press is a department of the University of Oxford. It furthers the University's objective of excellence in research, scholarship, and education by publishing worldwide. Oxford is a registered trade mark of Oxford University Press in the UK and in certain other countries

© Oxford University Press 2020

Authors: Justin Woolliscroft and Katy Patchwood

The moral rights of the authors have been asserted

Database right Oxford University Press (maker)

First published 2006

New editions published 2008 and 2014

This edition published 2020

All rights reserved. No part of this publication may be reproduced, stored in a retrieval system, or transmitted, in any form or by any means, without the prior permission in writing of Oxford University Press, or as expressly permitted by law, by licence or under terms agreed with the appropriate reprographics rights organization. Enquiries concerning reproduction outside the scope of the above should be sent to the Rights Department, Oxford University Press, at the address above.

You must not circulate this work in any other form and you must impose this same condition on any acquirer

British Library Cataloguing in Publication Data
Data available

978-0-19-848986-3

13

Paper used in the production of this book is a natural, recyclable product made from wood grown in sustainable forests.
The manufacturing process conforms to the environmental regulations of the country of origin.

Printed in Great Britain by Ashford Colour Ltd, Gosport.

Acknowledgements

The publisher and authors would like to thank the following for permission to use photographs and other copyright material:

Cover: OUP/Shutterstock. **p5**: Midland Aerial Pictures/Alamy Stock Photo; **p6**: Shutterstock; **p17**: Granger/Shutterstock; **p29**: Arch White/Alamy Stock Photo; **p36**: Courtesy of Dundee Satellite Receiving Station; **p43**: David Harlow/Getty Images; **p44**: NASA; **p54**: ZUMA Press, Inc./Alamy Stock Photo; **p57**: Sean Pavone/Alamy Stock Photo.

Artwork by Kamae Design, Mike Phillips, Steve Evans, Ian West, Giorgio Bacchin, NAF, Ruth Palmer.

Every effort has been made to contact copyright holders of material reproduced in this book. Any omissions will be rectified in subsequent printings if notice is given to the publisher.

The manufacturer's authorised representative in the EU for product safety is Oxford University Press España S.A. of El Parque Empresarial San Fernando de Henares, Avenida de Castilla, 2 – 28830 Madrid (www.oup.es/en or product.safety@oup.com). OUP España S.A. also acts as importer into Spain of products made by the manufacturer.

Contents

1 Fieldwork, and GIS — 4
1.1 The fieldwork that changed the world — 4
1.2 What kind of fieldwork will you do? — 5
1.3 What are the stages in fieldwork? — 6
1.4 A sample fieldwork report — 7
1.5 What is GIS? — 8
1.6 GIS in fighting crime — 9

2 Population — 10
2.1 How is Earth's population changing? — 10
2.2 So where is everyone? — 11
2.3 Population growth around the world — 12
2.4 How is the UK's population changing? — 13
2.5 What is our impact on our planet? — 14
2.6 What does the future hold? — 15

3 Urbanisation — 16
3.1 How did our towns and cities grow? — 16
3.2 Manchester's story – part 1 — 17
3.3 Manchester's story – part 2 — 18
3.4 Urbanisation around the world — 19
3.5 Push and pull factors — 20
3.6 It's not all sunshine! — 21
3.7 Life in the slums — 22
3.8 How can we make cities more sustainable? — 23

4 Coasts — 24
4.1 What causes waves and tides? — 24
4.2 What work do the waves do? — 25
4.3 Which landforms do the waves create? — 26
4.4 What do we use the coast for? — 27
4.5 Your holiday in Newquay — 28
4.6 Storm surge! — 29
4.7 How long can Happisburgh hang on? — 30
4.8 How can we protect places from the sea? — 31

5 Weather and climate — 32
5.1 Weather: what, why, and where? — 32
5.2 How is heat carried around Earth? — 33
5.3 Air pressure and our weather — 34
5.4 Why is our weather so changeable? — 35
5.5 What's a depression? — 36
5.6 More about rain … and clouds — 37
5.7 What's a tropical cyclone? — 38
5.8 Climate and climate factors — 39
5.9 So what's the UK's climate like? — 40
5.10 Climates around the world — 41

6 Climate change — 42
6.1 Earth's climate – always changing! — 42
6.2 The climate detectives — 43
6.3 How is Earth's climate changing today? — 44
6.4 This time … is it us? — 45
6.5 Local actions, global effects — 46
6.6 What can we do? — 47

7 Asia — 48
7.1 What and where is Asia? — 48
7.2 Asia's countries and regions — 49
7.3 What's Asia like? — 50
7.4 What are Asia's main physical features? — 51
7.5 Asia's population — 52
7.6 Asia's biomes — 53

8 China — 54
8.1 China: an overview — 54
8.2 A little history — 55
8.3 Mainland China's physical geography — 56
8.4 Where is everyone? — 57
8.5 How Shenzhen became a megacity — 58
8.6 Life in rural China — 59
8.7 What about the environment? — 60
8.8 What's the Belt and Road Initiative? — 61

Fieldwork, and GIS

1.1 The fieldwork that changed the world

This is about Doctor Snow's very clever use of maps over 165 years ago.

1. Look at these statements. They describe what Doctor Snow did, but they are jumbled up. Write a number from 1–7 in each box to put them in the correct order.

   Work began on the sewage system in 1859. ☐

   Using a map of the area, Doctor Snow marked all the households where people had died. ☐

   On 31 August 1854, a cholera outbreak hit the area called Soho, in London. ☐

   Doctor Snow looked for patterns. ☐

   He also marked where the water pumps were. ☐

   Within ten days, 500 people were dead. ☐

   Doctor Snow thought that the water from the Broad Street pump was infected. ☐

2. Imagine it is 1857 and you are Doctor Snow. Write a letter to the government explaining why a new sewage system should be built. Use facts and figures to support your ideas.

   ....
   ....
   ....
   ....
   ....
   ....
   ....
   ....

Tip! Include details from Doctor Snow's fieldwork to help you persuade the government that you are right.

4 Fieldwork, and GIS

1.2 What kind of fieldwork will you do?

This is about the aims of fieldwork and how you can collect data.

1. In the town of Ironbridge, there are lots of gift shops and cafes for tourists, but not many shops selling things that local people need. Some students are preparing to carry out fieldwork about shopping facilities in the town.

   They have made a list of enquiry questions, but can't decide which one to use. You can help them by ranking these questions in order of how suitable they would be for geography fieldwork. Justify your decision for each one.

   Rank

   ☐ How many shops are in this town?

   Justification: _____

   ☐ To what extent do the shops meet the needs of local people?

   Justification: _____

   ☐ Do local people use the shops?

   Justification: _____

2. After deciding on an enquiry question to focus your fieldwork, the next step is to collect data. There are different types of data: quantitative, qualitative, primary, and secondary.

   Here is a list of different ways to collect data. Can you add them to the table? Look back at page 8 of *geog.2* for a reminder of the different types of data if you get stuck.

   Looking for census data online
   Drawing a field sketch
   Studying newspaper articles
   Counting pedestrians
   Comparing land use on map
   A tally chart of plant species
   Interviewing people
   Finding old photos
   Measuring the length of rocks
   Taking photos
   Measuring distance on a map

Quantitative		Qualitative	
Primary	Secondary	Primary	Secondary

Fieldwork, and GIS

1.3 What are the stages in fieldwork?

This is about the stages in fieldwork and how you can stay safe when collecting data.

1. Geography fieldwork is divided into stages.
   Use arrows to link these stages up with their explanation. The first one has been done for you.

   Stage 1: Enquiry question or hypothesis — This can be counting, measuring or asking questions.

   Stage 2: Plan your fieldwork — At this stage, you should explain what have you found out.

   Stage 3: Collect the data — The aim of fieldwork is to answer this question or test this hypothesis.

   Stage 4: Process and present the data — Careful consideration of what you will do and how you will do it.

   Stage 5: Analyse the data — Comment on what went well and how you could improve any part of your enquiry.

   Stage 6: Draw conclusions — You can do calculations, draw graphs and annotate photos.

   Stage 7: Evaluate your enquiry — Write about the patterns and trends. Suggest reasons.

2. As part of the fieldwork planning stage, students need to think about how they will stay safe. Annotate this photo of a river to show potential dangers for students collecting data there.

3. Now explain how these risks could be avoided.

   ....

   ....

   ....

1.4 A sample fieldwork report

This is an opportunity to think about fieldwork you could do in your school.

1 A map is a useful starting point for fieldwork. In the box below, sketch a map of your school like the one on page 13 of *geog.2*. Include the outside areas, because fieldwork is more likely to be done in the school grounds than inside the buildings. Draw in pencil so that you can make changes.

Have you included?
- [ ] playground areas
- [ ] fields
- [ ] sports facilities
- [ ] car parks
- [ ] outdoor seating

2 One type of fieldwork you could carry out is a **microclimate enquiry**.

A microclimate is the climate of a small area. For example, one side of a playground that is always in the shade is likely to be cooler than the other side, which catches the sunshine. As part of a microclimate enquiry, you could measure the temperature and wind speed at different locations around your school to find out if there are differences.

a What equipment might you need? Complete these sentences.

A thermometer measures ....

An anemometer measures ....

b Now think of five places around your school grounds that you could investigate. Add these to your sketch map, using numbers 1–5 for each location.

3 What are your predictions? Which places will be warmest/coolest? Which places will be windiest? Include reasons for your predictions.

Fieldwork, and GIS

1.5 What is GIS?

The world is increasingly data rich. GIS helps us to effectively use data to solve problems.

1. GIS needs data. Fill in the gaps in the following passage using words from the box beneath. Use the student book to help you if you get stuck.

   GIS displays _____ on a map, helping us to find _____ , make _____

   and decide on _____ to take. GIS data is very well-organised in _____ .

   Each layer has one _____ . We can turn layers on and off, so that we are only seeing

   the layers that are _____ to us. All of the data is tagged with the _____ and

   longitude of the place where it was collected.

theme	decisions	helpful	actions	layers	data	patterns	latitude

2. Look at these data layers for a new GIS investigation. People were asked in a survey if they would like to see a new shopping centre built.

   Describe one investigation that this information would allow you to do. Explain how each of the data layers would help you.

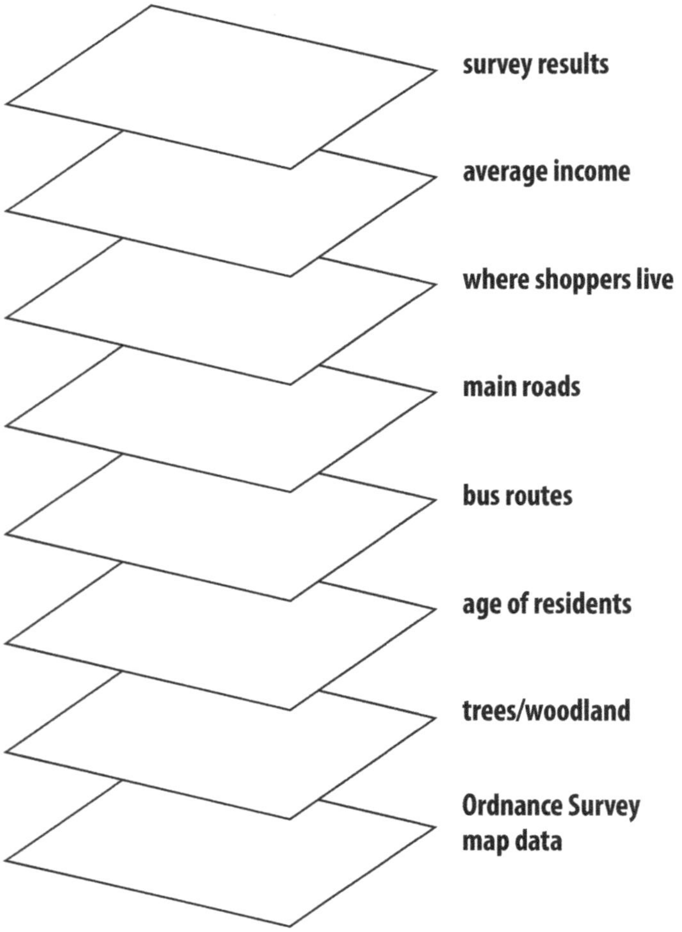

   - survey results _____
   - average income _____
   - where shoppers live _____
   - main roads _____
   - bus routes _____
   - age of residents _____
   - trees/woodland _____
   - Ordnance Survey map data _____

Fieldwork, and GIS

1.6 GIS in fighting crime

The police do lots of fieldwork, and use GIS to help them solve crimes.

1 Imagine you live in a town suffering from a high number of household break-ins. How might GIS help you to work out where the burglar will strike next? Explain your thinking.

....
....
....
....

2 The map on page 16 of *geog.2* shows where crimes were committed. Look at the crimes recorded in grid square 1137. Write in the blog template below giving advice to people visiting the area. The blog post has been started for you.

Home	Archive	About	🔍

Keeping safe

When visiting this area it is really important that you keep safe by...

Tip!
Be specific. Look at the types of crime and think about how people could avoid becoming crime victims.

3 You are in charge of controlling crime in the area. Where would you put most of your police resources? Name specific locations from the map and give reasons for your answer.

....
....
....
....
....

Population

2.1 How is Earth's population changing?

This is about population change, and how quickly our numbers are rising.

1. Use arrows to link the key terms below to their correct definitions.

   birth rate   birth rate minus death rate

   natural increase   the number of deaths each year for every 1000 people

   death rate   the number of births each year for every 1000 people

2. **a** In the UK, life expectancy is 81 years. Edna lives in Yorkshire and is 103 years old. Explain how this is possible.

   b Edna had a younger sister, Mary, who died as a baby, about 100 years ago. Today, far more babies survive to be adults. Why?

3. Write down two factors that you think would reduce the death rate. For each, explain why.

Factor	Explanation

4. Imagine you are a farmer in 1200 BCE. You've just bought some brand new iron tools to use. How might they have an impact on the population of your village in years to come?

Tip! Think about how better tools would affect the amount of food you can grow.

2.2 So where is everyone?

This is about countries with lots of people, and places with hardly any.

1. We call countries with lots of people densely populated, and countries with few people s _____ populated.

2. Look at the map. You will notice that Greenland and India have been left blank. Use the key to shade in the country that you think is densely populated, and the country you think is sparsely populated.

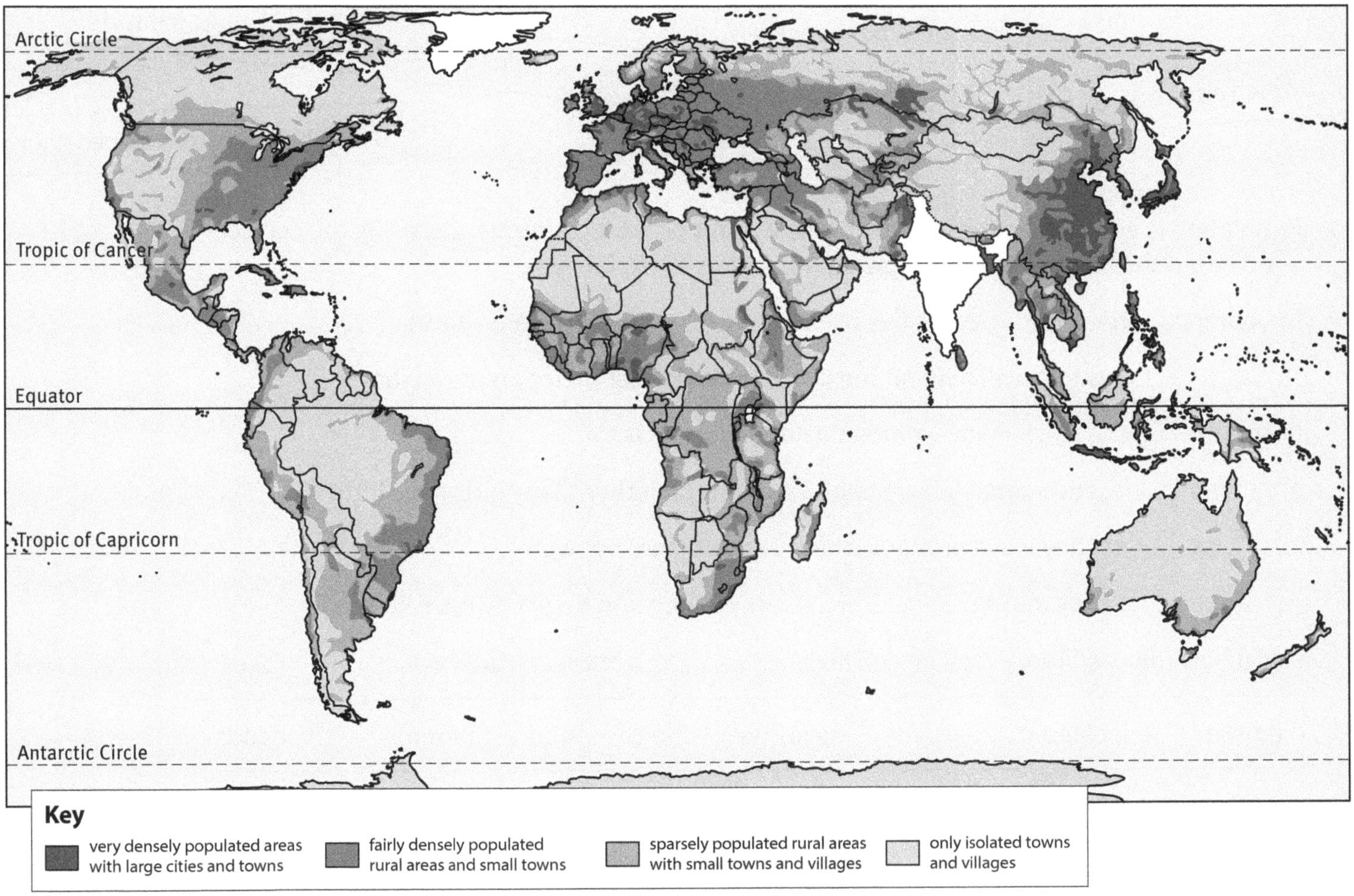

Key
- very densely populated areas with large cities and towns
- fairly densely populated rural areas and small towns
- sparsely populated rural areas with small towns and villages
- only isolated towns and villages

3. Look at the map again, and decide which of the following statements are true and which are false. Circle **true** or **false** beside each statement.

   a The northern hemisphere is more crowded than the southern hemisphere. **true / false**

   b Places near the poles are more densely populated than places near the tropics. **true / false**

   c Africa is more densely populated than Europe. **true / false**

   d Australia is more sparsely populated than Japan. **true / false**

   e The population of South America is evenly distributed across the continent. **true / false**

Population

2.3 Population growth around the world

This is about population growth and fertility rates around the world.

1 Draw lines to match the heads and tails of these sentences.

- Population is rising fastest in Africa, …
- The Earth's human population is growing at about 1.1% per year…
- Population growth is generally faster…
- …which is the poorest continent.
- …the northern hemisphere.
- The countries where the population is falling are found in…
- …which adds over 80 million people to Earth each year.
- …in poorer, less-developed countries.

2 Population is growing fastest in some of the world's poorest countries. Choosing the correct words from the box, complete the paragraph below to explain why.

In poorer countries, most people live by _____. Children are a form of _____, as they will help on the farm and look after their parents as they become _____. Many women do not have access to _____ about family life, and so many have babies one after another. Many girls have little _____ as they leave primary school early and are not well _____. They may be _____ very young and they may have little say in how many children they have. Their _____ are in control.

| advice | educated | married | security | choice | old | farming | husbands |

3 In wealthier countries, fewer babies are born. Complete this table to explain why.

Factor	Explanation: why does this lead to fewer babies?
More women choose to work and have careers.	
Raising children can be expensive.	
Having a large family may harm the environment.	

12 Population

2.4 How is the UK's population changing?

This is about how the UK's population has changed over time.

1. The Industrial Revolution saw the construction of many factories. Imagine you are one of the child miners shown in the drawing on page 26 of *geog.2*. Write 50 words describing how you feel about your work.

   ....
   ....
   ....
   ....

   Tip! Remember that there may be good and bad points – try to think of both.

2. Since 1801, the government has held a census or population count every 10 years. Before this it didn't know exactly how many people lived in the UK. What problems do you think this may have caused for the government?

   ....
   ....
   ....

3. Using the table below, calculate which countries are experiencing natural increase and which are experiencing natural decrease. The UK has been done for you.

   Tip! Remember that to calculate natural increase, subtract the death rate from the birth rate.

Country	Birth rate	Death rate	Natural increase / decrease
Brazil	14	6	
Norway	11	8	
Nigeria	38	12	
UK	11	9	2 – increase
Ukraine	9	15	

   Source: World Bank

Population 13

2.5 What is our impact on our planet?

This is about our increasing use of resources and what the future may be like as a result.

1. Using the information from the student book, complete the spider diagram to show how the demand for the world's resources is growing. One has been done for you.

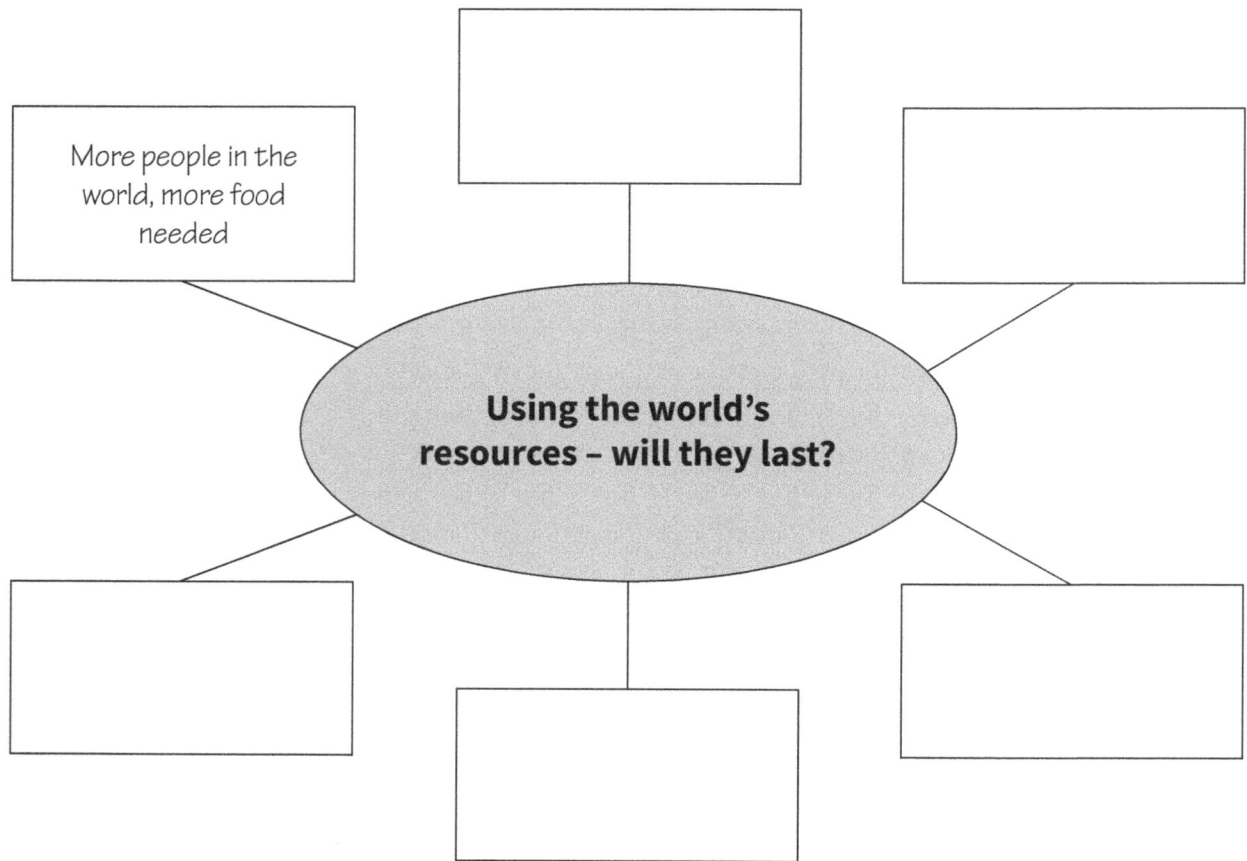

2. We can all help the planet by living in a more sustainable way. Here are some examples.

   | A Walk, don't drive | B Recycle and re-use |
   | C Don't use plastic bags | D Buy locally grown food |

   Choose one of the above and explain why you think it would help our planet.

   Example chosen: ....

   Explanation: ....

   ....

   ....

   ....

   ....

14 Population

2.6 What does the future hold?

This is about using evidence to plan for a growth in population.

1. The UK's population may reach 76 million by the year 2045 – that is an extra 9 million people living in the UK compared to today. How do you think this may affect our lives? Add words to the speech bubbles below to explain the effects of this population growth on each of the people and their jobs.

Hospital doctor

Lorry driver

Teacher

National Park warden

2. The United Nations chose 11 July as World Population Day, to help raise awareness of the challenges and opportunities that a rising world population brings. Former Secretary-General, Ban Ki-moon, gave this message for World Population Day on 11 July 2014.

   "On this World Population Day, I call on all with influence to prioritise youth in development plans, strengthen partnerships with youth-led organisations, and involve young people in all decisions that affect them. By empowering today's youth, we will lay the groundwork for a more sustainable future for generations to come."

   Do you think that young people can provide the answer to halting the growth in world population? Give reasons for your answer.

   ....
   ....
   ....
   ....

Urbanisation

3.1 How did our towns and cities grow?

This is about the reasons behind the growth in world urbanisation.

1. These sentences describe how our towns and cities grew, but they are in the wrong order. Put them into the correct order by writing a number from 1 to 7 in the correct box.

   The Industrial Revolution meant factories were built near towns so they could get workers. ☐

   Some villages grew into market towns. ☐

   Towns became bigger and bigger, and some became cities. ☐

   Clusters of dwellings became settlements. ☐

   New farm machinery, built in factories, meant that not so many farm workers were needed. ☐

   Villages grew around markets. ☐

   Farm workers moved to towns to find work in factories. ☐

2. What are the consequences of rapid urban growth? Write your answers in the hexagons below. Two examples have been done for you. Try to be balanced – both positive and negative ideas!

   Rapid urbanisation means that...

   - there will be more job opportunities.
   - more people will need houses.

3. Now write down three links between the outer hexagons. For example, more job opportunities may mean that people can afford to rent or buy housing. Compare your answers with a partner.

   ....
   ....
   ....

3.2 Manchester's story – part 1

This is about the urbanisation of Manchester.

1. Study the photograph below. In the spaces around the photo, write down one thing that you know about the photograph, two things that you don't know and three things that you would like to know more about.

I know …

I dont know …

I dont know …

I would like to know …

I would like to know …

I would like to know …

2. Imagine you lived in Manchester at this time, and you worked in a mill. Write a diary entry about what you think your typical working day would be like. Try to use the words below in your answer.

| tired | safety | painful | poor |

Urbanisation 17

3.3 Manchester's story – part 2

This is about the regeneration of Manchester.

1 Below is a diagram showing the two key aims of regeneration. Add labels around the edge to show some examples of each aim. Then compare with the completed diagram on page 38 of *geog.2*. Were there any you missed? Add them in a different colour.

Tip! Remember that some examples fit into both aims.

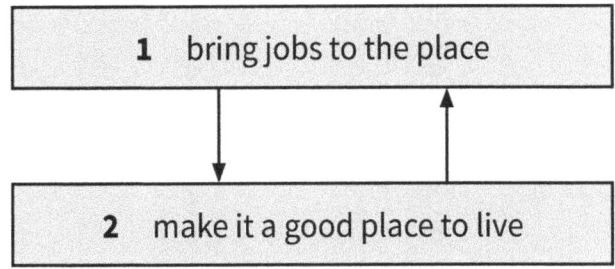

The aims of regeneration

1. bring jobs to the place
2. make it a good place to live

2 Welcome to MediaCityUK, built on the site of the docks where much of Manchester's historic trade was carried out. Now it is a waterfront destination for Greater Manchester, with digital creativity, learning and leisure at its heart.

The BBC and ITV both operate at MediaCityUK, producing thousands of hours of content for television, radio and online.

The University of Salford also operates at MediaCityUK and says "it is a vibrant place in which to live, work, socialise and study."

"MediaCity UK truly represents Manchester's future."

How far do you agree with this statement? Give reasons to explain your answer.

....
....
....
....

Urbanisation

3.4 Urbanisation around the world

This is your chance to find out about urbanisation in other countries around the world.

A megacity is a city with over 10 million people. In 2019, there were 33 megacities in the world. This number is likely to keep growing.

Rank	City	2019 population	Estimated Annual Growth Rate	Estimated Population in 2030	Potential Natural Hazard/s
1	Lagos, Nigeria	21 million	3.34%	18.9 million	None
2	Dhaka, Bangladesh	20.3 million	3.56%	28.1 million	Cyclone Drought Flooding
3	Shenzhen, China	12.1 million	1.88%	14.5 million	Cyclone Drought Flooding
4	Karachi, Pakistan	15.8 million	2.24%	20.4 million	Cyclone Drought Flooding
5	Delhi, India	29.4 million	3.03%	38.9 million	Drought Flooding

1 Most of the cities in the table above are also at risk from natural hazards. Use an atlas to locate and label the five cities on the world map below. Then write at least three sentences to explain why the expected population growth in these cities may cause problems.

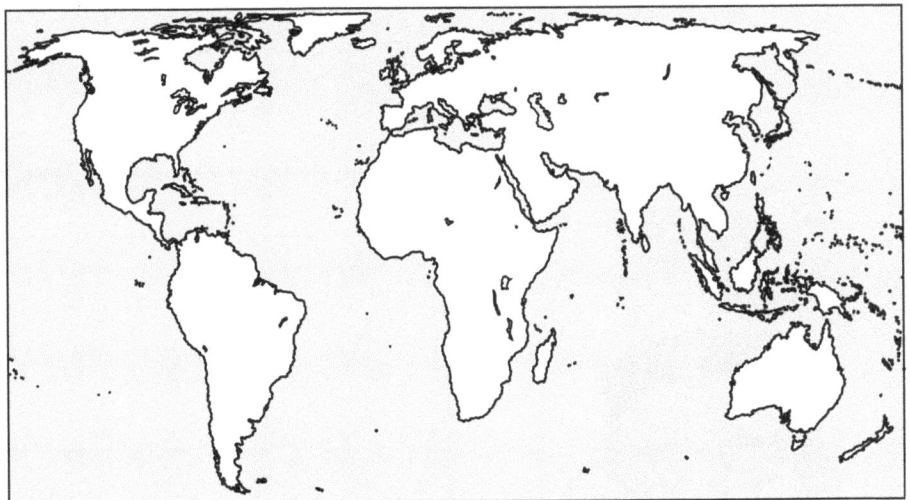

Tip!
Think about the impacts on the environment, people, and the economy of the country and city.

....
....
....

Urbanisation 19

3.5 Push and pull factors

This is where you will discover the reasons why people move from rural to urban areas.

1. Think about your own life. Write down three reasons why you might choose to move to another town, city or country in the future. For each reason, explain whether it would be caused by a push or pull factor.

	Reason for moving	Push / Pull	Explanation
1			
2			
3			

2. For each of your own reasons above, describe one change to your own area that would make you more likely to stay.

Urbanisation

3.6 It's not all sunshine!

This is where you will discover the pros and cons of city life.

1. Page 44 of the student book shows some of the benefits and disadvantages of urban living. Choose any three of each and list them below.

Benefits of urban living	Disadvantages of urban living

2. In the space below, design an area of a city where you would like to live. It should have each of your chosen benefits, but also should have some design features that mean the disadvantages are no longer a problem!

Tip! Think as creatively as you can and give your city an appropriate name!

Urbanisation 21

3.7 Life in the slums

This is where you will find out about the poorest people who live in cities.

1. The paragraph below describes life in the slums.

   Circle the correct word from each pair. Use page 46 in the student book to help you.

   Cities are growing **quickly** / **slowly** in developing countries. Many slums are made from anything the people can find, and have **no** / **some** running water. Osakwe lives in a slum in **Lisbon** / **Lagos**. Nine people live in **two** / **four** rooms, so it is very crowded. There is a lot of rubbish, and people throw it into the **ditch** / **bins** outside the house. In Lagos, around **one-third** / **two-thirds** of the population live in slums.

2. Look at the photographs of slums on pages 46 and 47. Choose three adjectives to describe what they are like. The first should have 4 letters, the next should have 6, and the last should have 8 letters.

   Adjective 1 (4 letters) _____   **Adjective 2** (6 letters) _____

   Adjective 3 (8 letters) _____

3. One way to tackle the slum problem is through self-help schemes, and many charities support these. Save the Children, a British charity set up in 1919, also says that:

   'Education is many children's route out of poverty. It gives them a chance to gain valuable knowledge and skills, and to improve their lives. And it means when they grow up, their children will have a much better chance of surviving and thriving.'

   If you were making the decision on how to spend £5 million to help people like Osakwe in the slums of Lagos, how would you spend it? Write a letter to the Nigerian government explaining your decision.

   > **Tip!**
   > You can use page 47 of the student book for ideas, but remember to explain your decision.

3.8 How can we make cities more sustainable?

This is where you can display your creativity!

1 You have been asked to design the world's most sustainable city. Fill in the table below to show how it will be sustainable. You can use drawings, diagrams, and writing. Be creative!

Transport	Housing
Waste	Environment
Food	Water

Urbanisation

Coasts

4.1 What causes waves and tides?

This is about what causes the waves and tides, and the different types of waves.

1 Choose words from the box to fill in the gaps in the sentences below.

sun	weak	boats	wind	moon	big	small	long
backwash		short	fetch	carry		strong	swash

Waves are made by .... pulling on the surface of the water.

The length of water over which the wind blows is called the ....

Large waves are made by:

- .... wind
- the wind blowing for a .... time
- a .... fetch

The water that goes up the beach when a wave breaks is called the ....

The water that goes back down the beach is called the ....

Tides are caused by the pull of the ....

2 Write the correct caption from the bullet list below underneath each diagram of a wave.

- If a wave is high and steep, it erodes the beach.
- If a wave is low and flat, it erodes the beach.
- If a wave is high and steep, it builds up the beach.
- If a wave is low and flat, it builds up the beach.

24 Coasts

4.2 What work do the waves do?

This is about what jobs waves do.

1. Write these phrases into the correct part of the table below.

   - the process is called longshore drift
   - how the waves wear away the coast
   - a beach is made like this
   - when waves drop the load they are carrying
   - low flat waves drop material
   - material is moved in a zig-zag by the swash and the backwash
   - this process involves solution, hydraulic action, attrition and abrasion
   - when material is carried along the coast

erosion	transport	deposition

2. Which process do you think needs the **most** energy? Why?

   ....
   ....
   ....

3. Which process do you think needs the **least** energy? Why?

   ....
   ....
   ....

Coasts

4.3 Which landforms do the waves create?

This is about how coastal landforms are made.

1 The diagram below shows several coastal landforms. Add the labels from the box to it.

```
crack
cave
stack
wave-cut platform
```

2 Explain how waves have created the landforms in the diagram.

....

....

....

3 What do you think will happen to these landforms next? Why?

....

....

....

Draw a diagram of your prediction here:

26 Coasts

4.4 What do we use the coast for?

This is about how the coast is used in many ways by different groups of people.

1. **a** People use the coast in many different ways. In the box below, write down as many examples as you can.

   b Now consider whether these uses are environmental, economic, or social. Choose three colours (one for each) and circle your answers. Some might be circled more than once!

2. Choose one of your examples of how we use the coast. Write a few sentences about how it might have positive or negative impacts on the coast. Some uses might have both!

Coasts

4.5 Your holiday in Newquay

This is about using an OS map to find out more about Newquay.

1. In the spaces below, fill in details for an information leaflet for tourists who want to visit Newquay. Look at the OS map on page 61 of *geog.2* to help you.

Newquay in Cornwall
A holiday for all ages!

Beaches
_____
_____
_____

Countryside
_____
_____
_____

Things to do
_____
_____
_____

2. You could travel to Newquay by road, rail or air.
   Choose how you would travel to Newquay and explain why.
   Is your chosen method sustainable?

   I would travel to Newquay by **road** / **rail** / **air**. (Circle your choice.)
   I would travel this way because:

   ....
   ....
   ....

28 Coasts

4.6 Storm surge!

This is about how the floods of December 2013 had a major impact on the country.

1 What caused the storm surge of December 2013? Circle the correct answer in each pair to complete this explanation.

A **low** / **high** pressure system was passing over the UK.

This caused the sea to **rise upwards** / **sink downwards** a little.

At that time of the month, the tide was at its **lowest** / **highest**, adding to the water level. Lastly, **gentle** / **strong** winds whipped up the waves and pushed them towards the coast.

2 Choose one of the photos on page 62 of *geog.2*. Now imagine that you lived there during the floods of December 2013. Write an account of what happened. You can be creative – but remember to use details that really happened at that time.

4.7 How long can Happisburgh hang on?

This is about how one village is falling into the sea!

1. The boxes on the right explain why Happisburgh is falling into the sea. But they are in the wrong order! Write the correct order in the circles, using the numbers 1–5.

2. This diagram shows what is happening. Label these features:

   groynes   revetments   cliff   houses at risk   beach

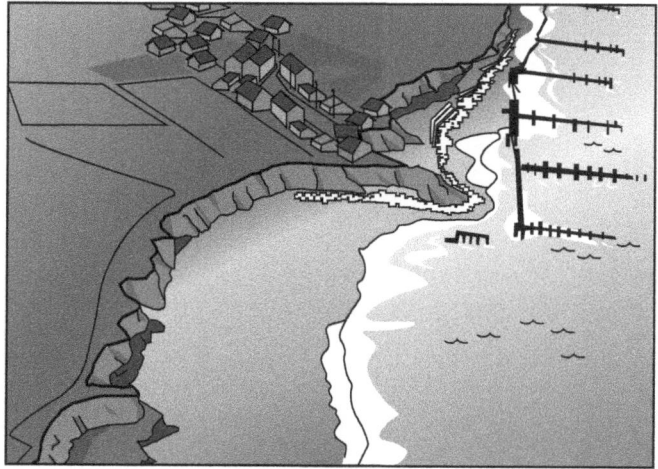

> The clay slides out of the bottom of the cliff and the sand on top collapses.

> All the time, the sea is also taking chunks out of the bottom of the cliffs by wave erosion.

> The cliffs are sand on top of clay below.

> The clay gets wet and the water makes it slippery.

> Rain can get through the sand to the clay.

3. The people whose homes are destroyed can't get the money from insurance. Cliff falls are called 'Acts of God' and aren't covered by insurance.

   a. How do you think the people feel about this? Why?

   b. Do you think they should get compensation? Why / why not? (Think about everyone else's insurance premiums, the amount taxes might go up, whether it's fair…)

Coasts

4.8 How can we protect places from the sea?

This is about how planners are trying to prevent erosion at the coast.

1 The pictures below show ways to reduce erosion.

   a For each one, write a title at the top. Choose from:
      Sea wall, Artificial reef, Beach nourishment, Groynes, Revetments, Rock armour.

   b Underneath each picture, explain how the method works.

   c Which method do you think is most effective? Why?

   d Which one do you think is least effective? Why?

Coasts 31

Weather and climate

5.1 Weather: what, why, and where?

This is about the big ideas that explain the weather.

1. Tick the correct answers to finish these sentences.

   a. As the Earth is round, it warms up…

   evenly ☐   unevenly ☐   slowly ☐

   b. To warm the Earth, the Sun sends out beams of…

   gravity ☐   vapour ☐   energy ☐

   c. At the Equator, it is warmer because the sunbeams are…

   bigger ☐   more spread out ☐   more concentrated ☐

   d. At the poles, it is cooler because the sunbeams are…

   smaller ☐   more spread out ☐   more concentrated ☐

2. Heat moves from warmer to colder places, which evens out the temperatures on Earth.

   Fill in the gaps to complete the statements. Choose words from the list below.

   | precipitation | water | sinks | wind | air | condenses |

   As it rises, it cools and _____ to form clouds, which are made of _____ droplets.

   _____ falls from clouds, in the form of rain, snow or sleet.

   In hot areas, warm _____ rises.

   The cooler air _____.

   Cool air is drawn in to replace the rising air. This is the _____.

   Tip! To help remember what happens when warm air rises, use the 3 C's: air <u>c</u>ools, <u>c</u>ondenses and forms <u>c</u>louds.

5.2 How is heat carried around Earth?

This is about air pressure and the global atmospheric circulation.

1. Air pressure affects weather. Low pressure air is light and it rises, cooling and condensing to form clouds – which bring rain. High pressure air is heavy and sinks, which means no clouds can form – creating dry, bright conditions.

   Now let's work on retaining this information! Draw an image in each box to help you remember the weather conditions associated with high and low pressure.

Low pressure	High pressure

2. Read the speech bubbles below. Fill in the gaps.

   You can use words such as: high, low, hot, cold, dry, sunny, wet, snowy.

   I live in Siberia in Russia, at a latitude of 60° N. There is _____ pressure and the weather is often _____.

   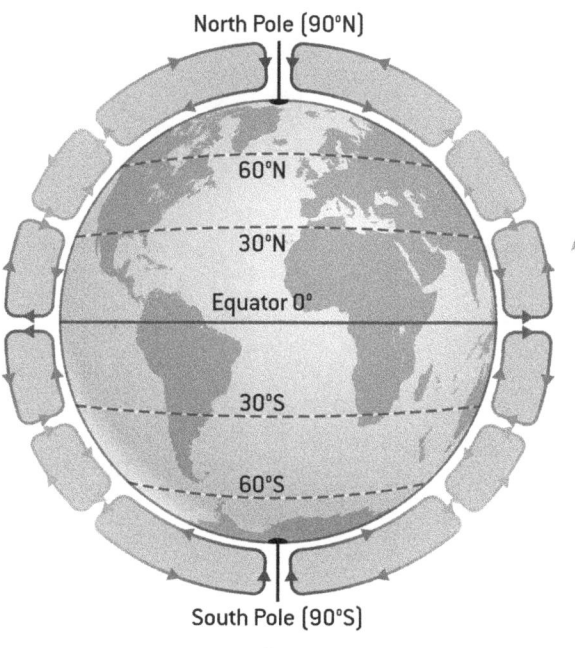

   I live at the Equator in the Brazilian rainforest. There is _____ pressure and the weather is usually _____.

   I live in the Namib Desert in Africa, at a latitude of 30° N. There is _____ pressure and the weather is usually _____.

   I'm a scientist working near the South Pole. There is _____ pressure and the weather is usually _____.

Weather and climate

5.3 Air pressure and our weather

This is about the weather you get with high and low air pressure.

1 The pictures show low and high pressure weather. Write the annotations below in the correct box on each picture.

> But warm rising air means clouds form …
>
> At B, the air pressure is higher. Air rushes from B to A as wind.
>
> At A, warm air is rising. The air pressure falls.
>
> … and clouds lead to rain.

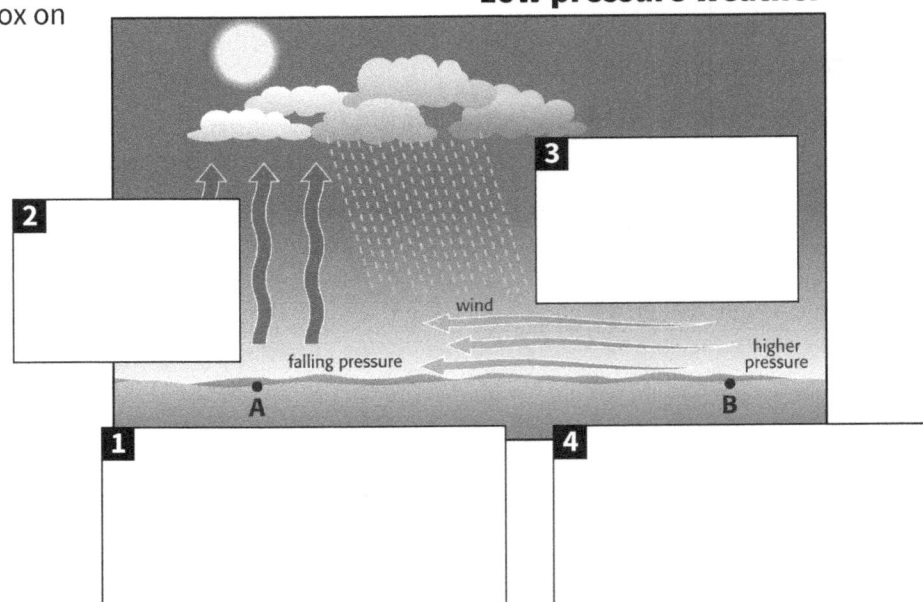

> The warm air cools, and sinks. Air pressure at Y rises.
>
> … so cold air gets pushed aside …
>
> Now Y has high pressure.
>
> Warm air is rising at X …
>
> As the cold air sinks, it warms up. So no clouds form over Y. The sky stays clear.

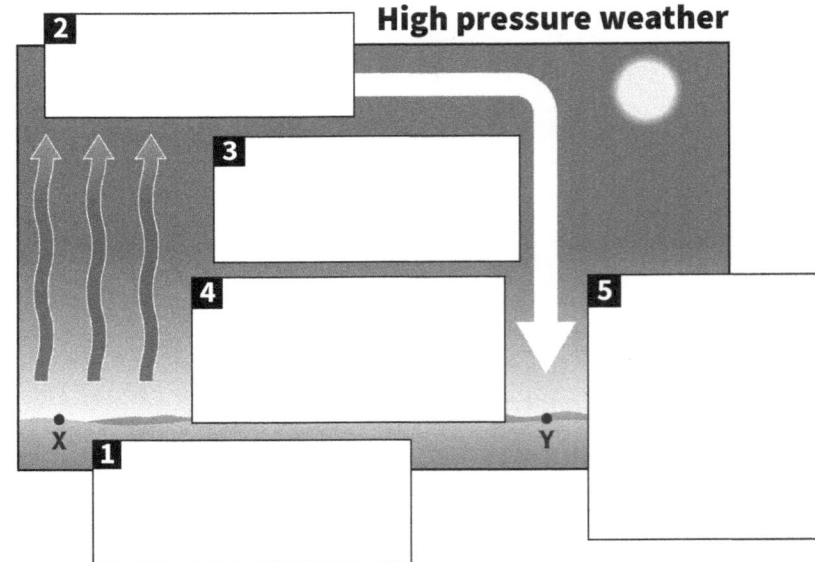

2 Describe the weather you get with high pressure in winter and summer. You need to mention whether there are any clouds, whether it is hot or cold, and whether there is any rain. Make sure you also include the words in the box.

| frost   dew   thunderstorms   fog |
| drought   flooding |

....

....

....

34 Weather and climate

5.4 Why is our weather so changeable?

This is about why our weather in the UK changes so quickly.

1 Are these statements true or false? Put a tick in the correct box.

		True	False
a	The air moves around the world in huge blocks called air masses.	☐	☐
b	A warm air mass brings strong gusty wind and heavy rain.	☐	☐
c	An air mass coming from the North Pole will be warm and damp.	☐	☐
d	The leading edge of an air mass is called a front.	☐	☐
e	A cold air mass brings wind and rain.	☐	☐
f	When a new air mass reaches the UK, it brings a change in the weather.	☐	☐
g	An air mass coming from a warm ocean will be cold and dry.	☐	☐
h	Warm air always moves from a warmer place to a colder one.	☐	☐
i	The UK is closer to the Equator than to the north pole.	☐	☐

2 The diagram on the right shows what happens when a warm air mass arrives. The text below explains what happens, but it is in the wrong order. Put it in the right order by writing the numbers 1–4 in the boxes.

☐ The rising air cools. The water vapour condenses to form a sloping bank of cloud.

☐ Warm air is lighter. So it slides up over the cold air.

☐ It starts to rain. It may rain for hours.

☐ As it rises, the pressure falls. So the weather gets a bit windy.

3 Draw the weather symbol for:

a A warm front

b A cold front

Finish the sentences.

A warm front means ....

A cold front means ....

Weather and climate

5.5 What's a depression?

This is about the weather conditions associated with depressions.

1 A depression forms when a warm, moist air mass meets a much colder air mass. The warm air mass pushes into the cold air mass. Look at this satellite image of a depression. From above, the warm air mass is shaped like a wedge.

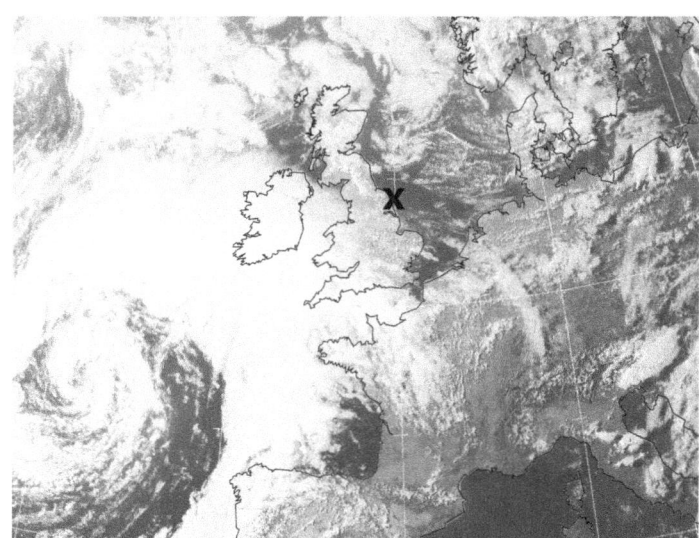

On the satellite image:

a Draw the warm front – a line with red semi-circles.

b Draw the cold front – a line with blue triangles.

c In front of the warm front, the air is cold. Lightly shade it, using blue.

d Behind the warm front, the air is warm. Lightly shade the wedge of warm air, using red.

e Behind the cold front, the air is cold. Lightly shade it, using blue.

2 Imagine you live at **X**, on the north east coast of England. A depression is heading your way!

The statements below describe what will happen as the depression passes over you, but they are jumbled up.

Write a number from 1–6 in the boxes to put them in the correct order.

☐ The depression has died away now. The weather has settled down.

☐ The cold front is overhead and there are big, towering clouds. You can hear thunder and it's pouring with rain. Get inside, quick!

☐ The depression has not arrived yet. The cold air mass is overhead. It's pretty chilly. Brrr.

☐ The warm front has passed over now, and you are in the warm air mass. You can take your coat off!

☐ The cold front has passed over and the air is cooler and but there are still a few rain clouds.

☐ The warm front is overhead. It's cloudy and starting to rain. Put up your umbrella.

3 Depressions sometimes bring very stormy conditions to the UK. In 2019, the organisers of several music festivals decided to cancel their events due to extreme weather warnings. Suggest reasons why the stormy conditions made it too dangerous to hold a festival.

....

....

....

Weather and climate

5.6 More about rain ... and clouds

This is about three types of rainfall.

1 These pictures show three different types of rainfall. Write the text beside each picture in the correct box on the picture.

Convectional rainfall

The rising air cools. The water vapour condenses. Clouds form. It rains.

The sun warms the ground ... which then warms the air above it.

Currents of warm air rise.

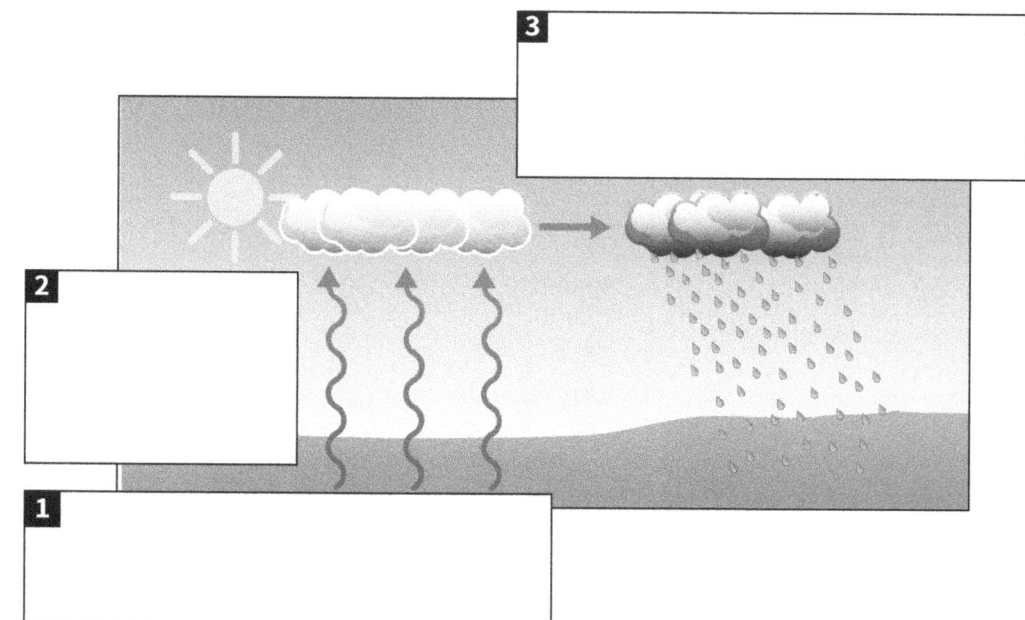

Relief rainfall

The rising air cools. The water vapour condenses. Clouds form. It rains.

Warm moist air arrives from the Atlantic Ocean.

The rain falls on the windward side of the mountain. The leeward side stays dry.

The air is forced to rise.

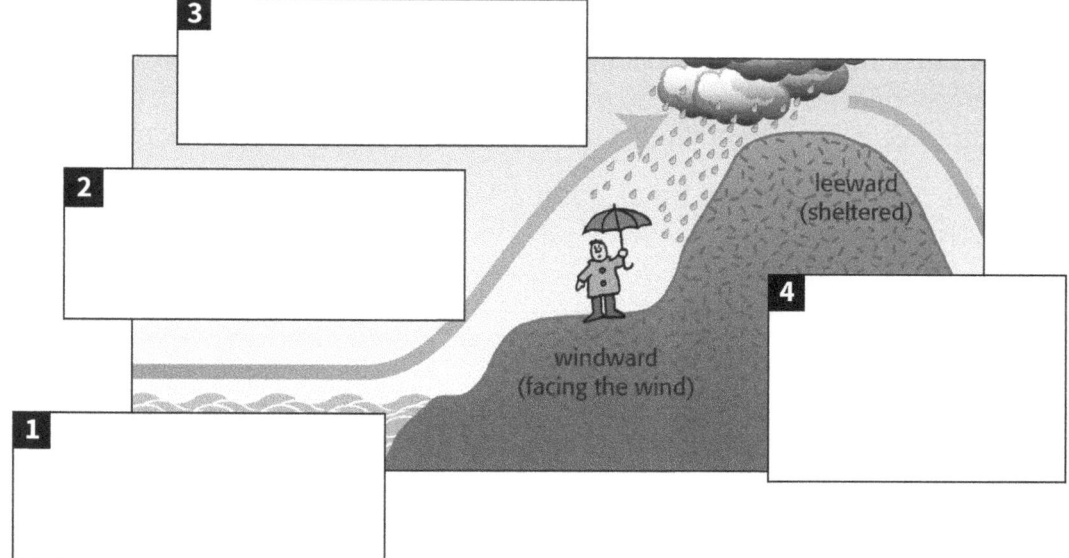

Frontal rainfall

The warm air mass slides up over the cold one, or gets driven up by it.

A warm air mass meets a cold air mass.

The rising air cools. The water vapour condenses. Clouds form. It rains.

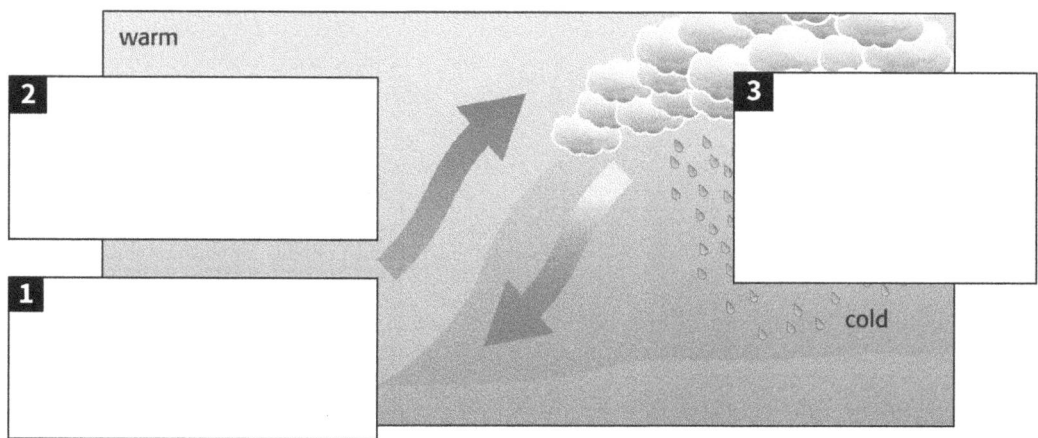

Weather and climate 37

5.7 What's a tropical cyclone?

This is about the distribution of tropical cyclones and how people can respond to them.

1. Study this map, which shows the distribution of tropical cyclones around the world. The arrows show the direction that the storms move. Using a world map or an atlas to help you, list five countries that are at risk of tropical cyclones in each of these continents:

   | North America |
   | Africa |
   | Asia |
   | Oceania |

   Key
   → Storm paths

2. Tropical cyclones can cause a lot of damage. People may lose their homes, their jobs, or even their lives. Buildings and infrastructure may be knocked down, blown away or flooded. Water supplies and power may be lost. In the aftermath, emergency support is required. In the weeks, months and years that follow, everything needs to be rebuilt.

   In the box below, make a list of jobs that would need to be done following a deadly tropical storm.

   Jobs to be done immediately:

   Jobs to be done over the next year:

I'm here to help!

38 Weather and climate

5.8 Climate and climate factors

This is about why climate is different in different places.

1 Draw a line to match each factor affecting climate with the correct effect below.

Factors	Effects
Latitude	In the UK, the North Atlantic drift warms the west coast in winter.
Distance from the coast	The further you go from the Equator, the cooler it gets.
Prevailing wind direction	A sea breeze keeps the coast cool in summer and warm in winter.
Ocean currents	In the UK, it's from the south west and brings rain.
Height above sea level/altitude	The higher you are above sea level, the cooler it is.

2 Look at the map and answer the questions. Use the examples in question 1 and pages 84–85 of the student book to help you.

   a Why is Margate always warmer than Aviemore?

   ....

   ....

   b Why will Leicester be warmer than Margate in summer, but cooler in winter?

   ....

   ....

   c Why is Penzance wetter than Margate?

   ....

   ....

   d Why is Oban warmer than Aviemore in winter?

   ....

   ....

Weather and climate

5.9 So what's the UK's climate like?

This is about the UK's climate and climate graphs.

1. Look at this map of the British Isles. We are surrounded by the sea! Prevailing winds and a warm ocean current come from the south west.

   Write down two ways that the sea affects the climate of the British Isles.

   ....

   ....

   ....

2. On a climate graph, a line is used to show temperature and bars are used to show precipitation.

   Look at this incomplete climate graph. The line shows average temperatures for Luton throughout the year. You can read the temperature values (in °C) from the left y axis.

   Now look at the right y axis, which shows rainfall (in mm). Finish the climate graph by drawing the bars to show rainfall. Use the data from the table.

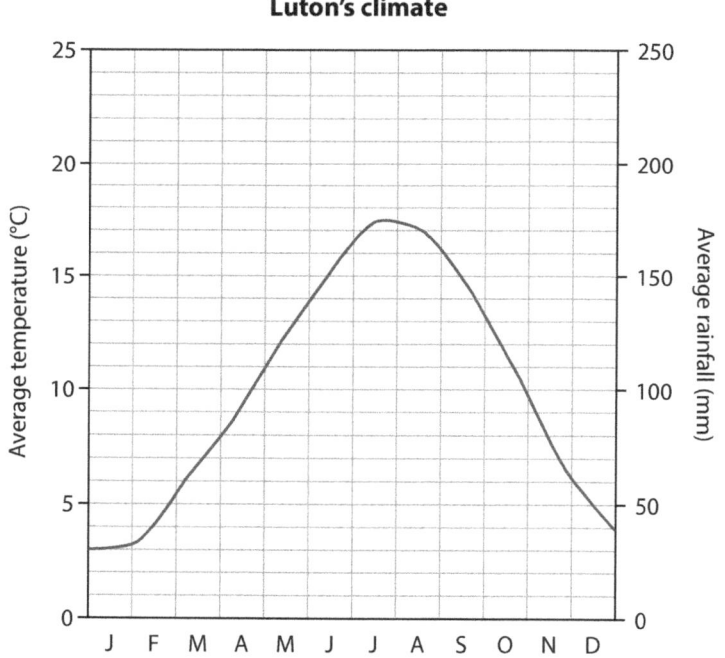

Month	Rainfall (mm)
Jan	59
Feb	41
Mar	51
Apr	51
May	52
Jun	55
Jul	49
Aug	55
Sep	57
Oct	59
Nov	59
Dec	63

3. Using the graph, you can describe the climate of Luton. Fill in the gaps.

   In Luton, ... is the hottest month of the year.

   The coldest month is ....

   .... is the wettest month of the year. The driest month is ....

   The total annual rainfall for Luton is ....

   The mean annual temperature of Luton is ....

Tip!
The mean is an average. To calculate the mean annual temperature, add up all the values and divide your answer by the number of months. You can use a calculator!

5.10 Climates around the world

This is about climate in different parts of the world.

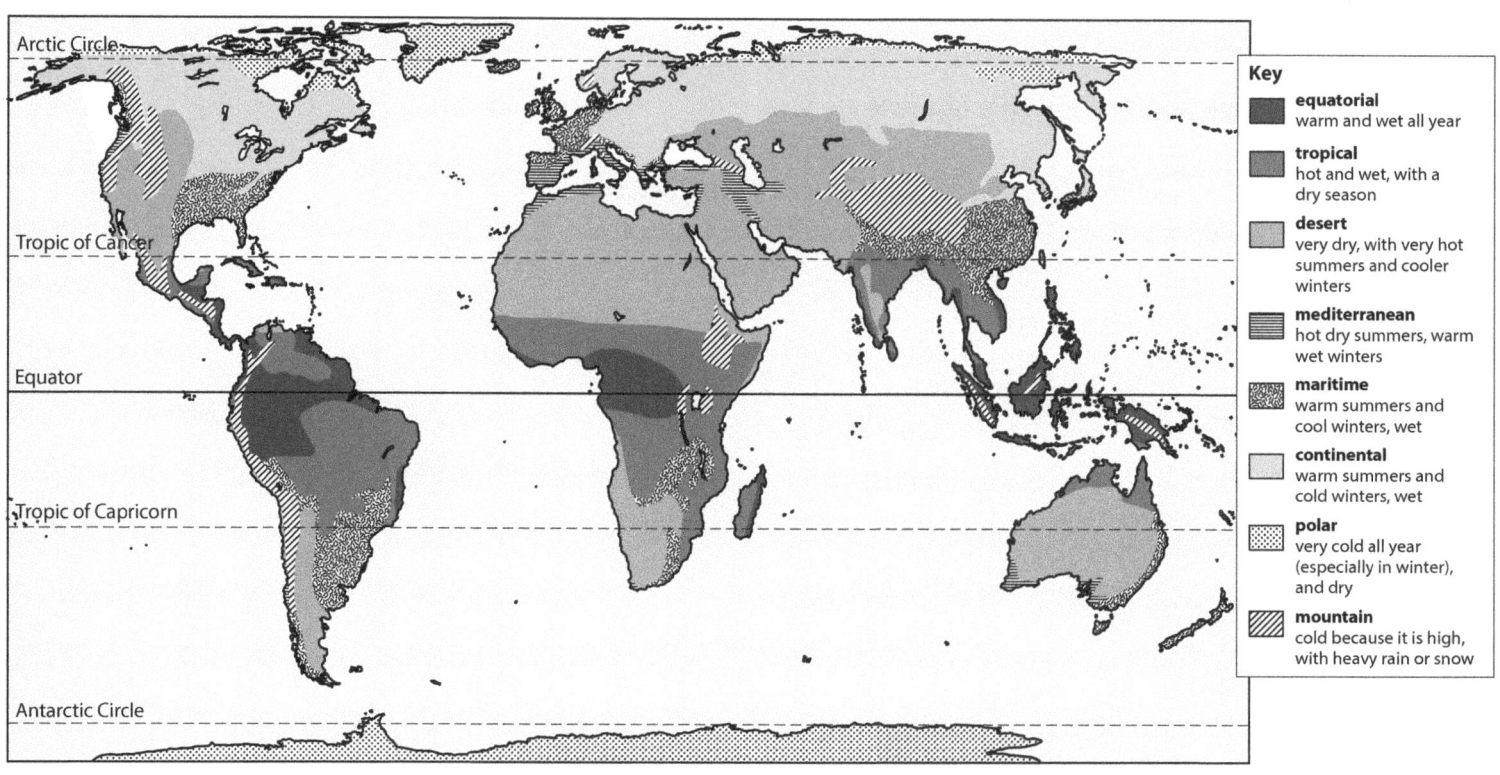

1. The map shows how the climate varies across the globe. Choose one of the climates you would like to go on holiday to. Explain why.

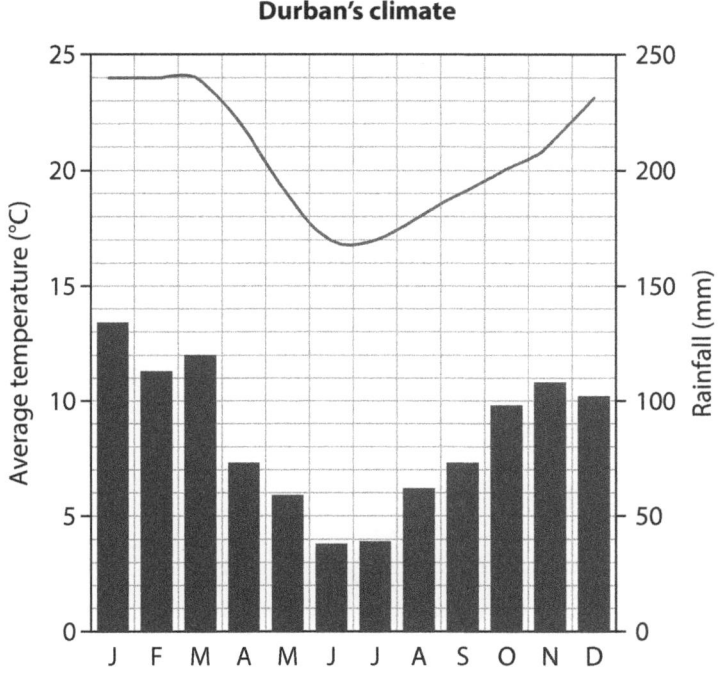

Name of climate ....

....

....

2. Look at this climate graph for Durban in South Africa. Compare the climate of Durban with that of Swansea. Use the graph on page 87 of the student book to help you and use numbers/figures from the tables on page 86 to back up your answer.

Temperature ....

....

....

Rainfall ....

....

....

Weather and climate 41

Climate change

6.1 Earth's climate – always changing!

This is about ice ages and the Earth's climate.

1 During the last ice age, sea levels fell by 120 m. Look at this map, which shows Europe – 20,000 years ago. Study the key carefully, then complete the sentences using the place names in the box.

At **A**, a 'land bridge' existed between .... and ....

At **B**, a 'land bridge' existed between .... and ....

.... and .... were completely covered in ice.

The region of .... was also under ice.

There was no ice in ....

North of ...., there was a huge lake.

If you tried to walk from .... to France, you would have had to cross over ice.

| Scotland | Great Britain | the Netherlands | Ireland | London |
| Scandinavia | the British Isles | Spain | Wales | France |

2 Why do you think sea levels fall during an ice age?

....

....

Tip! Think about rain, snow and rivers!

3 If Earth had another ice age, people living in the coldest places would either need to migrate or adapt to the changes.

a Can you think of any ways people could adapt?

....

....

....

b What problems could result from mass migration? Think of at least two.

....

....

....

....

6.2 The climate detectives

This is your chance to be a climate detective!

1. When Mount Pinatubo in the Philippines erupted in 1991, the blast was so large it affected the global climate for two years, making it colder. As a climate detective, your mission is to find out whether this variation in climate affected the trees in an area of woodland close to your school.

   Think about:
   - How many trees will you test?
   - Should you consider different species of trees?
   - Are there any trees that would not be useful for this investigation?

   Enquiry question: **Did the 1991 Mount Pinatubo eruption affect trees in my local woodland area?**

   How will you carry out this mission? Write your answer in the box below.

   Tip! Remember, you can drill a plug from a living tree, so you don't have to cut any trees down!

2. Mount Pinatubo's eruption affected the climate for two years. Supervolcanoes are volcanoes that are much larger than normal volcanoes – some are thousands of times larger than Mount Pinatubo. There is one in Yellowstone Park in the USA. Make a prediction about how a supervolcano eruption would affect the global climate.

   ....

   ....

   ....

Climate change

6.3 How is Earth's climate changing today?

This is about the impacts caused by Earth's changing climate.

1. In the albedo effect, light colours reflect sunlight (which keeps them cool) and dark colours absorb it (which makes them warmer).

   Look at the image of Greenland in winter. As Earth's temperatures rise, ice is melting in places like Greenland, revealing the ground underneath. Can you explain how this may lead to further warming in Greenland?

   ....

   ....

   ....

2. Climate change is causing some places to get wetter, while other places are getting drier. In the boxes below, explain how floods and droughts could affect people and the environment.

Floods	Droughts
People	People
Environment	Environment

3. Imagine you have been asked to speak about climate change during an assembly at a local primary school. Write a speech about why climate change matters, explaining the impacts that are already occurring. Include your ideas from question 2, but don't forget to mention other impacts, such as heatwaves and wildfires!

....

....

....

....

....

Climate change

6.4 This time ... is it us?

This is about how humans are contributing to today's climate change.

The data proves it's mostly our fault.

We are damaging our planet for the future.

Some people say it's our fault, others say it is a natural event. I'm confused.

I need my car. How am I supposed to work without it?

1 Look at what the people are saying above. Give one reason why they have that point of view.

Scientist ....

....

Environmentalist ....

....

Pupil ....

....

Driver ....

....

2 Which one of the above do you most agree with? Circle your choice and explain why.

　　　　　Scientist　　　　Environmentalist　　　　Pupil　　　　Driver

....

....

....

....

Climate change

6.5 Local actions, global effects

This is about rising emissions of greenhouse gases and the impacts on poorer countries.

1. Anote lives in the island nation of Kiribati, in the Pacific Ocean. His father works as a fisherman. His mother looks after their house and garden. They grow most of their own food, generate their own electricity using solar panels and they don't have a car. They have never been on a foreign holiday.

   Most people in Kiribati do not contribute to greenhouse gas emissions – but climate change is destroying their lives. Most islands in Kiribati are no higher than two metres above sea level. Tidal waves regularly flood the villages and wash away houses and crops.

   Some people think that by 2050, these islands will have disappeared forever.

   If he were given the chance to speak to world leaders about climate change, what do you think Anote might say? Write your answer in the speech bubble.

2. Some countries are beginning to make changes and have reduced their carbon dioxide emissions. If we are to limit the rise to 1.5°C then drastic action is needed. The UK is now aiming for 'net zero' carbon by 2050.

   What do you think 'net zero' means? How would it help tackle climate change?

46 Climate change

6.6 What can we do?

This is about how we can slow down global warming.

These actions could help to stop global warming.

1 Give out free bikes to everyone

2 Breed plants that will gobble up carbon dioxide

3 Put big taxes on air travel

4 Build more wind farms, for electricity

5 Don't turn on the heating. Just put on warm clothes

6 Turn off all the town and city lights at night

7 Shut down all power stations that use oil, coal or gas

8 Allow homes to have electricity for only six hours a day

9 Find a way to bury carbon dioxide under the ocean

10 Pass a law that women can only have one child each

11 Shoot millions of small mirrors into space, to reflect some sunlight away

12 Ban international events like the Olympic games

1 Choose **two** of these actions and explain why they may have disadvantages.

Number [ ] ....

....

Number [ ] ....

....

2 Choose one of the actions that you think is the best. Write a letter to the Prime Minister explaining your view.

....

....

....

....

....

Climate change 47

Asia

7.1 What and where is Asia?

This is about locating the continent of Asia on the world map.

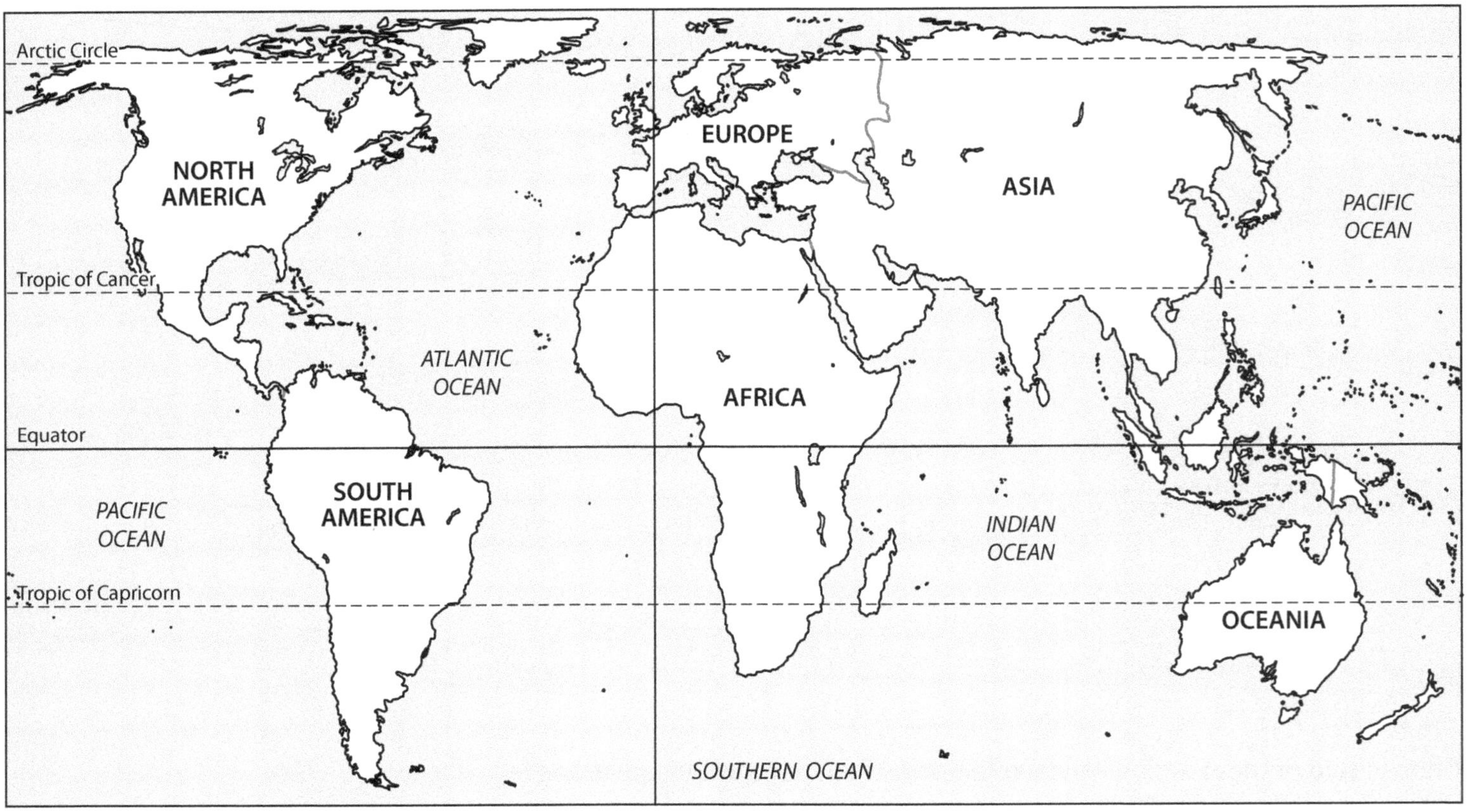

1 Colour in all the land in Asia that lies between the Tropic of Cancer and the Tropic of Capricorn. This area is known as the 'tropics' and is the hottest part of the world.

2 Using a second colour, colour in all the land in Asia that lies north of the Arctic Circle. This area is known as the Arctic. It is one of the world's two polar regions. They are the coldest parts of the world.

   a Where does Asia have most land – inside the tropics or inside the Arctic Circle?

   ....

   b Asia stretches from the tropics to the Arctic. What other continent stretches from the tropics to the Arctic?

   ....

   c Name the island that lies half in Asia, half in Oceania.

   ....

3 Using a third colour fill in all the land area of Asia that lies outside the tropics and outside the Arctic. What proportion of Asia lies neither within the tropics nor within the Arctic? Circle the correct answer.

   About 65%   About 80%   About 50%

4 The map also shows the Equator. Shade in the parts of Asia that lie south of the Equator.

48 Asia

7.2 Asia's countries and regions

This will help you understand Asia's size and the variety of its countries and regions.

Tip! Use the map on page 108 of *geog.2* to help you locate these capital cities.

1 Write down the names of the capital cities of the following countries:

   a China _____   b Saudi Arabia _____   c Sri Lanka _____

   b Iraq _____   e Cambodia _____   f Mongolia _____

   c Iran _____   h Kazakhstan _____   i Pakistan _____

2 Where are the Urals? You can find the Ural Mountains on the border between Europe and Asia on page 112 of the student book. Draw a line of triangles to show them on the map above.

3 Once you have drawn in the Urals on the map, guess which is the larger: China or Asian Russia?

   _____

4 Find the UK on the map above. What is the quickest way to get to the UK by sea from Japan? Why is this not possible for most of the year?

   _____

   _____

Asia 49

7.3 What's Asia like?

This will help to identify Asia's largest cities and put them into rank order.

Only 42% of Asia's population lives in urban areas. However, it has the second fastest urban population growth rate of any continent.

Many claims are made for the growth of Asia's megacities. The figures below are for the 20 largest cities in Asia, but the cities are already even bigger!

Tip! Remember that a megacity is a city with over 10 million people. In 2010, not all of these cities were megacities, but they have grown fast!

1 Place these cities into rank order and identify which country they are in.

City	Population (millions)	Country	Rank Order
Bangalore	12.33		
Bangkok	10.54		
Beijing	20.46		
Chennai	10.97		
Chongqing	15.87		
Delhi	30.29		
Dhaka	21.01		
Guangzhou	13.30		
Istanbul	15.19		
Jakarta	10.78		
Karachi	16.09		
Kolkata	14.85		
Lahore	12.64		
Manila	13.92		
Mumbai	20.41		
Osaka	19.17		
Shanghai	27.06		
Shenzhen	12.36		
Tianjin	13.59		
Tokyo	37.39		

2 It is very difficult to find agreement about the size of the world's largest cities. Can you think of reasons why population estimates vary?

50 Asia

7.4 What are Asia's main physical features?

Here we look at the relationship between Asia's physical features and its countries.

1. Compare the map on page 112 of the student book with the political map on page 108 and answer the following questions;

   a In which countries are the Himalayas?
   ....

   b In which countries does the Gobi Desert lie?
   ....

   c In which country is Lake Baykal?
   ....

   d In which country are the Zagros Mountains?
   ....

   e Name 5 rivers that rise on the Plateau of Tibet.
   ....
   ....

2. On the map:

   a circle the three groups of islands that are part of India.

   b circle the highest and lowest points in Asia.
      What is the difference in height between them? ....

   c circle the three islands named on the map that are part of the Philippines.

3. Research and find out what is unusual about the Aral Sea.

....
....
....
....
....

Asia 51

7.5 Asia's population

Here we look at the relationship between Asia's population and its physical features.

1. Environmental factors can help to explain population density. Choose two colours to complete the key below, then shade each box to show whether it is a feature of a densely or a sparsely populated place.

   1. poor soils and few plants
   2. no fresh water
   3. rivers with fresh water
   4. steep, cold mountains
   5. flat land for farming
   6. hot, dry desert
   7. plenty of rain
   8. fertile soils for crops

   Key
   ☐ densely populated
   ☐ sparsely populated

2. Look at the population density map on page 114 of the student book. In many parts of Asia, but especially in the north, the population distribution looks 'stripy'. Can you explain the reason for this?

   ....

   ....

3. Look up Indonesia on the political and physical maps of Asia on pages 108 and 114 of the student book.

   a. What do you notice about the population density of Java compared with the other islands of Indonesia?

   ....

   b. Can you explain this difference in population density?

   ....

   ....

4. The overall population density of Mongolia is 2 people per square kilometre, while the overall population density of China is 139 people per square kilometre. Explain why figures like this can be misleading. You should refer to the map on page 114 of the student book to help you explain your answer.

   ....

   ....

   ....

7.6 Asia's biomes

This is about telling what biome a place is in from its weather statistics.

Here are the climate charts for the five cities marked A–E on the map. Using the table, match each one to a numbered chart and write down which of the biomes it represents (tundra, mountain, temperate forest, hot desert, warm moist forest). Consider the height above sea level (elevation) of each place when looking at the temperature figures. Also fill in the reasons you chose each chart.

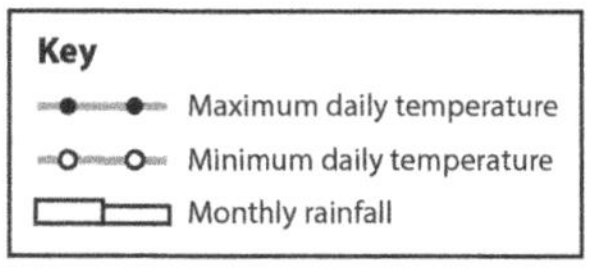

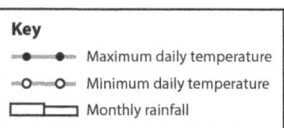

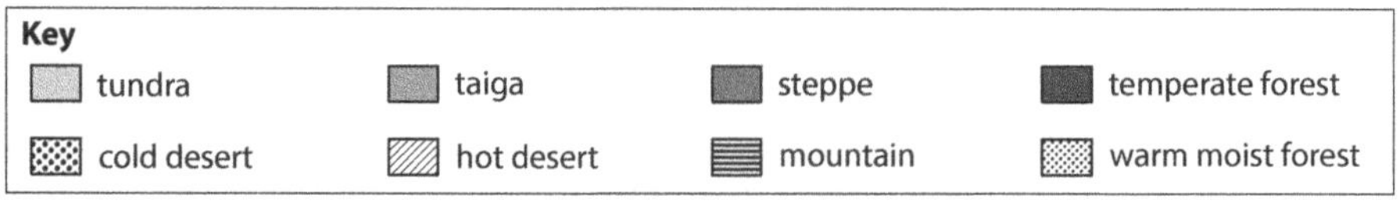

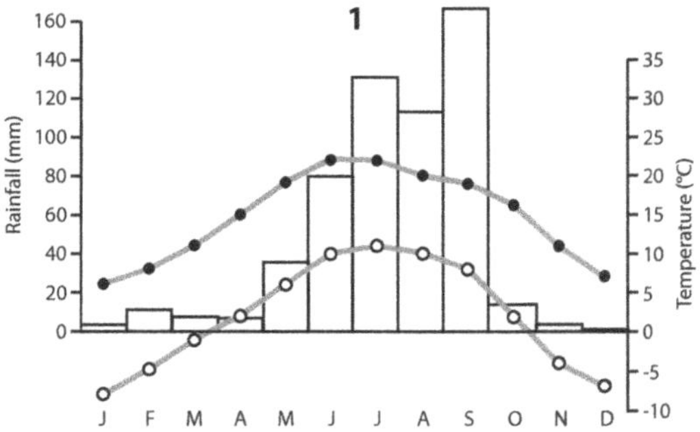

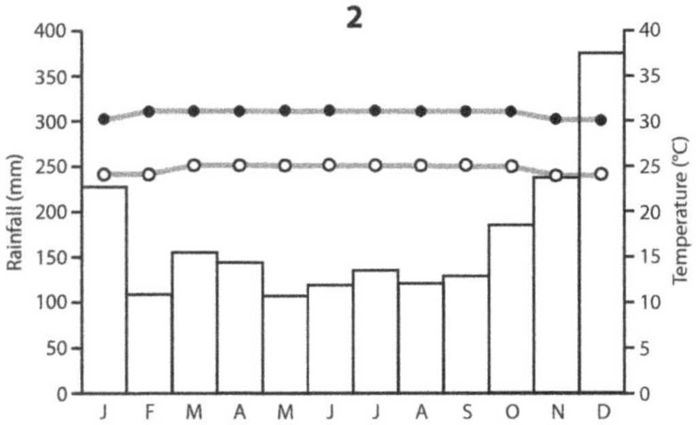

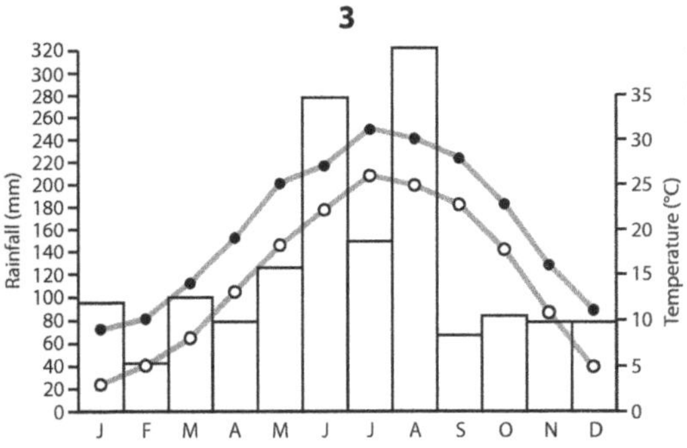

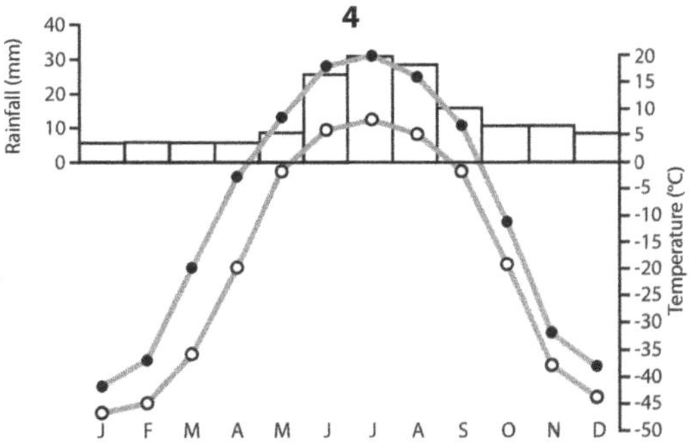

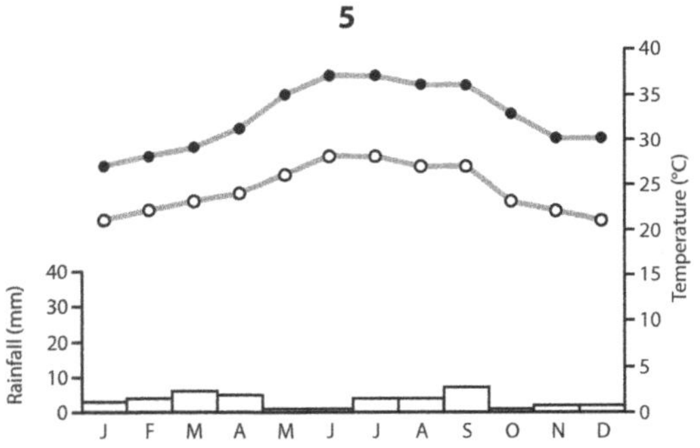

Key
— Maximum daily temperature
—o— Minimum daily temperature
▢ Monthly rainfall

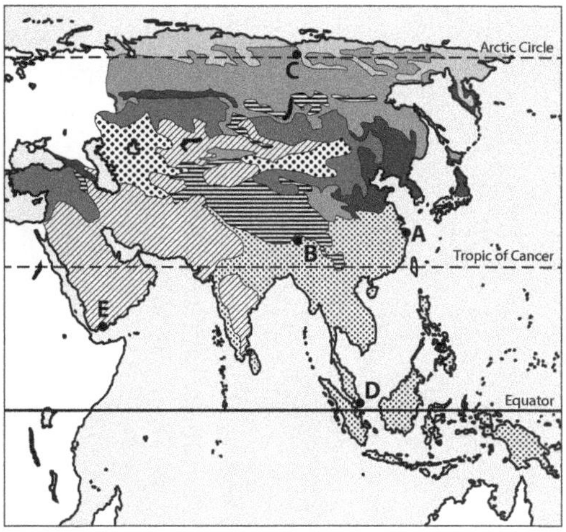

City	Elevation (m)	Biome	Chart number	Reasons
A: Shanghai	140			
B: Lhasa	3600			
C: Verkhoyansk	0			
D: Singapore	0			
E: Aden	0			

Asia 53

China

8.1 China: an overview

This is about China – a big important country.

1. Look at this map. China is a big country and shares its border with many neighbours.

   China's western neighbours have been added to the map. Here is a list of China's other neighbouring countries, marked on the map with the letters **A–I**. Use the clues to help you identify them and write the corresponding letter in each box.

   ☐ Vietnam – A long, thin country with a coastline in the South China Sea.

   ☐ Laos – Also long and thin, but it doesn't have a coast (it is 'landlocked').

   ☐ India – A big triangular country which will soon take over from China as the world's most populous country.

   ☐ Russia – A huge country in the north, which contains Siberia. Borders China in the east.

   ☐ Myanmar – Has a long coastline in the Indian Ocean.

   ☐ Nepal – A rectangular country. Contains Mount Everest!

   ☐ Bhutan – A small landlocked country, known for its strong cultural heritage.

   ☐ Mongolia – Cold, vast plains, deserts and mountains in the north.

   ☐ North Korea – A small country that shares its north-east border with China. It has a very strict government.

2. China is developing quickly and aims to be the world leader in artificial intelligence. Look at this police robot, pictured here in the streets of Beijing. What jobs do you think it can do?

   ....
   ....
   ....

54 China

8.2 A little history

This is about the changes that have occurred in China since 1949.

1. China was ruled by emperors for much of its recorded history. In the last 120 years, however, it has undergone enormous changes. The illustrations in the student book show its story from 221 BCE to 1949.

    Draw your own illustrations for these more recent key events in its history – from 1949 to the present day.

Tip! To pronounce 'Deng Xiaoping', say this: 'dang – show – ping'!

When the Communist party took over in 1949, it aimed to make everyone equal. It wanted China to be strong and self-sufficient in every way.

The state took over all the land, factories and businesses. People were told what work to do and in return they were provided with free food, education, and health care.

Overall, however, it was not a success. It was hard to grow enough to feed everyone. Between 1958 and 1961 there was a great famine. Over 20 million people died.

In 1978, a new leader called Deng Xiaoping made some big changes. He set up Special Economic Zones along the coast to attract foreign companies.

Many businesses were set up and thrived. Now workers could earn money. At first there were mostly clothing factories, but there is now a focus on high-tech innovation.

Today, China is still run by the Communist party, but it mixes its ideas with capitalism. By 2050, it is predicted that China will be the richest country in the world!

China

8.3 Mainland China's physical geography

This is about mainland China's physical features and its climate.

1. This paragraph describes China's relief and climate.

   Choose words from the box to fill the gaps.

   China is so diverse. It has flat _____, high _____, huge cities, a vast network of _____ and five different climate _____. China's relief is highest in the _____. The Plateau of Tibet is 4000m above _____ level and the _____ mountains are even higher – and very cold. In the _____, there are deserts which are _____ in summer but very cold in winter. In the _____, temperatures are milder and there is flat, fertile farmland on the _____ Plain.

   | north | Himalayan | Hubei | east | mountains |
   | plains | warm | rivers | west | sea | zones |

   Tip! You could use the maps on pages 124 and 126 of the student book to help you, or an atlas that shows more features of China.

2. Plan a trip around China in which you will experience lots of different places. Complete the itinerary by stating where you will go and writing about what you might see in each place. Try to plan a diverse adventure! Think about the different features you could see: deserts, rivers, mountains, cities, rural areas…

   <u>My Travel Itinerary</u>

   Week 1: Arrive in Beijing airport, in north-west China. In Beijing, I think I will see _____

   Next, I will visit _____. The landscape will be _____ and the weather is likely to be _____, although this will depend on the time of year.

   Week 2: _____

   Week 3: _____

   Week 4: _____

   Arrive in Shenzhen, in south west China. I expect to see _____.

   Fly home from Shenzhen airport.

8.4 Where is everyone?

This is about population density in mainland China.

1. Urban areas in China have a high population density.

   What does 'population density' mean?

   ....

2. China's population was once rising rapidly. Now it is only rising slowly because the government introduced a one-child policy in 1979. It was lifted in 2016.

   a What was the one-child policy and what effects did it have?

   ....

   ....

   ....

   b Can you explain why cities in China are still growing rapidly even though population growth has slowed down?

   ....

   ....

   ....

3. Rural areas in China have a low population density. Look at this photo of a rural village in China. Imagine you are there, looking down into the valley. Can you draw a field sketch in the box? Remember to include labels around the edge, such as: rice terraces, steep slopes, apartment blocks.

China 57

8.5 How Shenzhen became a megacity

This is about Shenzhen, a city in China.

1. Imagine you are a new migrant to Shenzhen. You have come from a rural area and have never been to a city before. What do you see, hear and smell when you arrive? Write your thoughts below.

2. You turn around a street corner and see a bus with no driver. You are shocked!

   You speak to a passer-by and they tell you that the bus is being driven robotically by a computer, and that it is very safe.

   What do you think about driverless buses? Write some pros and cons in the table below.

Pros	Cons

3. Many people in Shenzhen have had their faces scanned into the government computer system. This data can be used to issue fines if they get spotted crossing a road when the lights are red.

   Can you think of some other uses of facial recognition technology?

China

8.6 Life in rural China

This is about families left behind in rural China.

1. Some people in rural China choose to leave to find work in a city. Ten-year-old Chan's mum and dad left him to live with his gran when they went to find work in Chongqing city. You can read his thoughts about this on page 131 of the student book.

   Imagine what Chan's parents might say if you asked them about it. Fill these speech bubbles.

   > We left our village because....

   > We couldn't bring Chan with us because...

2. Every month, Chan's parents send remittances back to Chan and his gran.

   a What are remittance payments?

   b Why do you think Chan and his gran will need them?

China 59

8.7 What about the environment?

This is about energy use in China and analysing related data.

1 This table shows China's energy sources in 2008 and in 2018.

Energy source	Amount of energy provided by this source (Mtoe*) ...	
	in 2008	in 2018
coal	1609	1907
oil	385	641
gas	70	243
nuclear	15	67
renewables	151	416

* Mtoe: millions of tonnes of oil equivalent

a On the bar chart below, the data for coal has been plotted. Can you finish the graph by adding the other energy sources? You will need: a pencil, a ruler and a rubber.

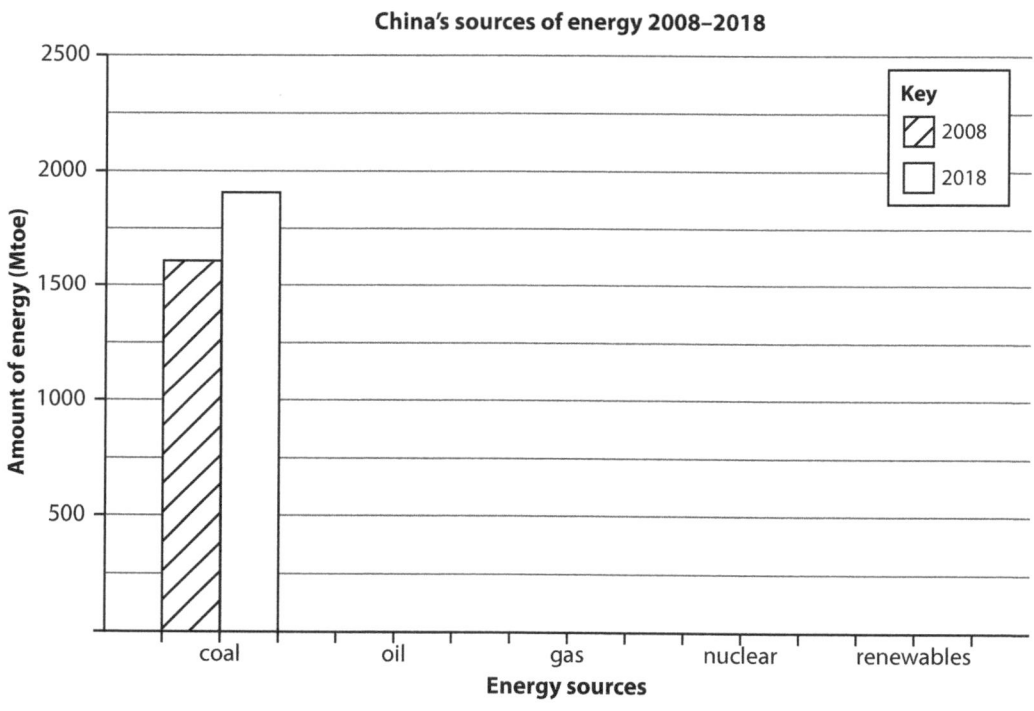

b What does the graph show? Fill in the gaps using the words from the box.

Between 2008 and _____, China continued to increase its use of fossil fuels. It uses more _____ than any other type of energy. However, cleaner _____ of energy are increasing too. Energy use from _____ sources more than doubled. 'Renewables' means things like _____ energy, _____ power and _____ electric power.

| wind | hydro | sources | renewable | coal | 2018 | solar |

60 China

8.8 What's the Belt and Road Initiative?

This is about China's Belt and Road Initiative (BRI).

1. The paragraph below describes China's Belt and Road Initiative (BRI). Circle the correct word from each pair.

   The BRI is a project to create a **small / huge** network of transport links from China across the world. It will make it **easier / harder** for goods to be imported and exported. The Belt is the links on **land / sea**. The Road is the links by **land / sea**. China announced it in **2013 / 2019**. It hopes to complete it by **2030 / 2049**.

2. Read these statements about China's Belt and Road Initiative.

   1 "The new port at Gwadar is great news for me and my family. I have got a well-paid job there, looking after the machinery."

   2 "Our government is repaying huge debts to China. This means there is less money to spend on public services. I work in the health service, so we are going to struggle."

   3 "A new railway has been built through my town. The trains are noisy, but I can now visit my sister who live 50 miles away."

   4 "It will be so much easier for me to transport my products once the BRI is completed. This means more profit for the company's shareholders."

   5 "China's BRI is a sign of its rapidly increasing wealth. Projections show that China will be the richest country in the world by 2050."

   6 "China's new ports are a dangerous threat to global stability. They could be used as military bases if war breaks out."

   a. Shade positive comments in green. Shade negative comments in red. For any comments that are both positive and negative, use red and green stripes.

   b. Match up the statements with this list of stakeholders, by writing the number of the matching statement next to each one.

   CEO of a coffee bean company ☐   Engineer at Gwadar port ☐
   Nurse in Sri Lanka ☐   Citizen of Hambantota, a town in Sri Lanka ☐
   United States government official ☐   University economics lecturer ☐

3. Now write another statement that could have been made by someone affected by the BRI. It could be by someone who lives in China or who lives in a country that the BRI routes pass through. Remember to include who said it!

   _____

   _____

your own notes...

geog.2 workbook

5th edition

This workbook is ideal for homework or independent study. It provides:
- engaging activities
- skills development
- support for every double-page spread in the geog.2 student book

geog.123 is a comprehensive course that matches the National Curriculum at Key Stage 3.

Did you know?
- Deserts are places with less than 25 cm of rain a year.
- They can be hot or cold!

What if...
- ... aliens abducted all the children?

The world's trusted geospatial partner

Also available in the series

Kerboodle provides online Lessons, Resources and Assessment to support teaching and learning in the classroom and at home. Online versions of the student book are also available.

geog.2 workbook – pack of 10
ISBN 978 0 19 848985 6

geog.2 workbook answer book
ISBN 978 0 19 848987 0

Printed on paper produced from sustainable forests.

OXFORD UNIVERSITY PRESS

How to get in touch:
web   www.oxfordsecondary.co.uk
email schools.enquiries.uk@oup.com
tel   +44 (0)1536 452620
fax   +44 (0)1865 313472

ISBN 978-0-19-848986-3